Headless WordPress, Formidable Power

An Enterprise Guide to Structured Data, APIs, and JavaScript Front Ends

Enterprise WordPress Developer Series

Book 1

Victor M. Font Jr.

Copyright and Licensing

Dedication

**To the developers who have outgrown templates,
and to the architects who remember when
WordPress was just a blog—
may your applications run structured, secure, and
boldly headless.**

Epigraph

"The tools we master today become the architecture of tomorrow's solutions."

Contents

Introduction

In the world of web development, many WordPress professionals reach a fork in the road. One path leads deeper into template tweaks, page builders, and shortcodes. The other leads toward structured data, API integration, business logic, and real-world application development.

This book is for those who choose the second path.

It's also the first volume in the Enterprise WordPress Developer Series—a collection of advanced technical guides designed to help you master scalable, secure, and structured development on the WordPress platform. Each book in the series stands alone, but together they form a roadmap for developers ready to build real applications, not just websites.

Whether you're a freelancer looking to elevate your practice or a developer seeking to transition into enterprise-grade systems, the move toward headless architecture is more than a trend—it's an invitation. It's an opportunity to decouple your thinking, modernize your workflows, and embrace a layered approach to solution design.

What follows is not a tutorial series, nor is it a dry theoretical volume. These chapters were born out of real conversations with real developers—especially those in the Formidable Masterminds community—who wanted more than drag-and-drop interfaces. They wanted frameworks. Patterns. Principles.

This guide delivers that.

We'll ride with a metaphorical headless horseman through the haunted woods of decoupled architectures, REST APIs, security gatekeeping, and frontend integration. Along the way, you'll find practical explanations, real-world code, and yes—a little dry humor where it counts.

You don't need to understand everything all at once. But by the end of this book, you'll not only know the path—you'll be ready to lead others along it.

Let's get started.

Preface

This book began as a series of thought experiments—what if WordPress wasn't just a website tool, but the backend of something greater?

As clients and enterprise systems increasingly demanded modularity, data integrity, and composable frontends, it became clear that headless architecture wasn't a trend—it was a trajectory. But what surprised me most was how WordPress—often underestimated—was already prepared for the journey.

The Headless Horseman is more than a metaphor. He represents velocity without fragility. Power without chaos. A backend without a predefined face.

And Formidable Forms? That's the secret weapon—capable of structured data modeling, workflow automation, and RESTful API interaction—all without abandoning the comfort and maturity of the WordPress ecosystem.

If you're a developer looking to move from builder to architect—or an enterprise decision-maker evaluating the future of your stack—this book will help you harness the potential of headless WordPress development.

Let's ride.

— Victor M. Font Jr.

Chapter 1

Enter the Headless Horseman: Why WordPress Decapitation Isn't a Horror Story

In the misty woods of Sleepy Hollow, the Headless Horseman rides—fast, focused, terrifyingly efficient. But while he may lack a head, he doesn't lack direction. And neither does headless WordPress.

Welcome, brave developers, as we ride boldly into the world of decoupled architecture—specifically, how to turn WordPress into a formidable (pun intended) enterprise back-end using

4

Formidable Forms, REST APIs, and your favorite JavaScript frameworks.

This is not a cautionary tale of haunted CMS choices. It's a roadmap for modernizing WordPress—without losing its soul.

What Does It Mean to Go Headless?

Let's clear up the fog. In a traditional WordPress site, the "head" is the presentation layer—PHP templates, theme files, and that glorious (or frustrating) mix of HTML and shortcodes that renders your site in-browser.

Going **headless** means lopping off that head and replacing it with something more agile—typically a JavaScript-based frontend like React, Vue, or Svelte. WordPress remains the back-end content engine, but the display is handled by a separate system via API calls.

Metaphor check: It's like giving our Horseman a motorbike instead of a horse. Still terrifyingly fast, but now optimized for terrain and speed.

Why Would Anyone Decapitate a Perfectly Functional CMS?

Three words: **scalability, speed, and flexibility**.

In an enterprise environment, you often need:

- Custom UIs tailored for devices and workflows
- Complex integrations with CRMs, ERP systems, or BI tools
- API-first architectures for SaaS-level modularity
- Frontends deployed separately from content systems

The monolithic WordPress stack—though versatile—is not well-suited for composable, multi-platform demands.

But with headless WordPress, you get the best of both:

- **Powerful content modeling** (CPTs, custom taxonomies, ACF, and Formidable Forms)
- **Enterprise-grade UX** via your favorite frontend stack
- **Separation of concerns** for better scalability and security

Where Formidable Forms Rides In

In our story, Formidable Forms plays the role of the powerful spellbook—the piece that empowers you to model and manage structured data without writing backend interfaces from scratch.

Here's why it's ideal in a headless setup:

- Entries are stored as structured meta fields (easily queried via REST API)
- Views can be disabled or bypassed entirely—no need for shortcode rendering
- You can use entries as custom API endpoints, or fetch them via the built-in REST route (/wp-json/frm/v2/forms/{id})

So whether you're building:

- A multi-stage onboarding wizard
- A complex data intake form for a regulatory system
- A frontend dashboard that visualizes user-submitted entries

...Formidable becomes your data wrangler—handling validation, field logic, repeaters, and calculated fields without burdening your frontend developers.

The API Awakens: How the Horseman Sees Without Eyes

WordPress' REST API allows any frontend to:

- **Pull data** (GET requests for entries, posts, users, taxonomies, etc.)
- **Push data** (POST requests for form submissions, new content)
- **Authenticate securely** (with tokens or cookie sessions)

And yes, Formidable's API is fully compatible with this model. Want to submit a form from your React frontend? You can use fetch() to POST directly to the REST endpoint. Want to retrieve a list of entries submitted by a logged-in user? A simple GET request will do.

Your "headless frontend" becomes a JavaScript-driven UI that communicates with the WordPress back-end over HTTPS. Clean. Modern. Secure.

Enterprise Use Case: Modernizing Legacy Intake Systems

Let's contextualize this with a common scenario. A client's legacy system needs structured intake, workflow approval, and role-based dashboards. WordPress and Formidable Forms offer rapid form modeling and relational data capabilities. But a decoupled frontend allows you to:

- Deploy a secure admin dashboard separately from the public site
- Customize the user interface without being constrained by theme architecture
- Integrate future microservices (e.g., notifications, mobile apps, analytics)

In short, you gain enterprise-level flexibility without abandoning the WordPress ecosystem.

Headless Wisdom

"When the head is cut off, the heart beats freer."
— Sleepy Hollow proverb (possibly made up)

Takeaways:

- Headless WordPress is not a novelty—it's a viable architecture for enterprise-grade applications.
- Formidable Forms is more than compatible—it's a strategic asset for modeling structured content.

- The REST API is your bridge—learn to cross it securely and wisely.

⚔

Next up: "*Anatomy of the Horseman: How the WordPress Body Keeps Moving Without Its Head*" — where we'll dissect how WordPress remains a powerful data engine without its native frontend.

Chapter 2
Anatomy of the Horseman: How the WordPress Body Keeps Moving Without Its Head

Last time, we introduced our darkly poetic metaphor: WordPress as the Headless Horseman—charging forward with purpose, even without a head (read: traditional front-end). Today, we dismount and dissect the anatomy of this mysterious rider.

What actually remains when you decouple the WordPress frontend?

The answer: quite a lot. While the themes and templates are gone, the engine—the **heart and spine** of your application—is very much alive.

☩

What Does the Headless "Body" Consist Of?

The decapitated WordPress still offers:

- Database schema: The wp_posts, wp_postmeta, wp_users, etc. tables remain your structured content core.
- Admin UI: Editors, authors, and admins still log into /wp-admin to create, manage, and publish content.
- Business logic: You still have plugins like Formidable Forms, ACF, custom post types, roles, workflows, and RESTful endpoints.
- REST API: /wp-json/wp/v2/ and other custom routes serve your content programmatically to the frontend.
- Authentication systems: Cookie auth, OAuth, or JWT tokens for securing interactions.

The headless body isn't brain-dead. It's a disciplined soldier, executing commands and processing input behind the curtain.

☩

Headless ≠ Useless Without Templates

Some developers wrongly assume that removing the frontend makes WordPress useless—like tossing a theme is equivalent to tossing the platform. Not so.

In a headless build:

- WordPress becomes your Content Repository or Backend-for-Frontend (BFF).

- Your API queries replace The Loop.

- Your JavaScript frontend handles rendering with frameworks like React, Vue, or Svelte.

- You gain decoupled freedom—deploying different frontends (web, mobile, kiosk) from the same data source.

Your new workflow looks more like:

- Author logs into WordPress and fills out a Formidable-powered data intake.

- Form stores structured data as meta fields.

- Frontend app (e.g., React) fetches that data via /wp-json/frm/v2/entries.

- UI renders entries in a custom-designed, performance-optimized layout.

It's not a haunted shell—it's a composable enterprise backend.

Core Components of the Headless Stack

Layer	Technology	Purpose
Data & Content	WordPress CPTs, Formidable Forms, taxonomies	Data modeling
Business Logic	Custom plugins, hooks, REST endpoints	Workflow and validation
APIs	WordPress REST API, Formidable Forms API	Data transport
Frontend	React, Vue, Svelte, or mobile app	UI rendering
Auth & Security	JWT, OAuth, cookies, WP nonce	Access control
Deployment	Netlify, Vercel, WP Engine Headless, Docker	Hosting and CI/CD

Formidable Forms: The Structured Soul

In a headless build, Formidable Forms can:

- Act as your admin-only UI for entering, editing, and managing structured content.

- Power calculated fields, repeaters, conditional logic, and entry relationships.

- Be extended to include custom REST endpoints for specialized workflows.

You no longer need Formidable Views for rendering content—your frontend handles display. But the form logic still lives in the body, enforcing integrity and business rules.

Real-World Architecture Sketch (Metaphorical)

Picture the Headless Horseman again:

- His spine is the WordPress database.
- His heart is Formidable Forms, pumping structured data through the system.
- His veins are REST APIs—transporting information to the head (the frontend).
- His armor is JWT/OAuth—securing the whole system.

This metaphorical anatomy holds up even at scale—whether you're building a multi-user SaaS app, a secure government portal, or an internal data dashboard.

⸎ ✝ ⸎

Headless Wisdom

> *"He doesn't need eyes to see. He has a route to follow."*

⸎ ✝ ⸎

Takeaways:

- Headless WordPress retains full back-end power—only the frontend is replaced.
- Your architecture now emphasizes APIs, data modeling, and frontend freedom.

- Formidable Forms plays a critical role in headless mode: structured input, validation, and admin UI.

Next up: "_The Reins and the Ride: React, Vue, or Svelte for the Frontend?_"—We'll explore how to choose the best frontend partner for your newly headless system—and what to consider when integrating with WordPress and Formidable Forms.

Chapter 3

The Reins and the Ride: React, Vue, or Svelte for the Frontend?

In our last two installments, we watched the WordPress Horseman lose his head—but not his horsepower. The backend remains alive and powerful, driven by APIs and structured by tools like Formidable Forms.

Now comes the question every headless architect must answer:

If WordPress is the horse...

What kind of frontend saddle do you want to ride on?

The headless ride demands a frontend framework worthy of the engine behind it. Today, we explore three modern stallions in the stable—**React**, **Vue**, and **Svelte**—each with its own gait, posture, and attitude.

Let's cinch up and ride.

⌐—┼—⌐

Why You Need a JavaScript Framework in the First Place

Traditional WordPress themes render UI with PHP. Once you go headless, your entire UI is driven by API responses. This means:

- You're fetching content from WordPress (often via REST or GraphQL)
- You're rendering interfaces dynamically using JavaScript
- You need state management, routing, and component lifecycles

This is why most developers turn to frontend frameworks like React, Vue, or Svelte.

Headless WordPress without a frontend framework is like a rider without reins—awkward, dangerous, and prone to getting lost in the trees.

✝

1. React: The Enterprise Stallion

Best For: Large-scale apps, rich component ecosystems, teams already familiar with JS tooling.

React is the industry favorite for headless WordPress projects, especially in enterprise environments. Why?

- Deep integration with frontend build tools (Webpack, Vite)
- Battle-tested state management (Redux, Zustand, React Query)
- Tons of prebuilt UI component libraries
- Used in platforms like Gatsby and Next.js (which offer WordPress integration)

WordPress Tie-Ins:

- Gutenberg is built with React
- WPGraphQL and headless starter themes often cater to React
- Easy to manage client-side routing and hydration from API-loaded data

Downsides:

- Steep learning curve
- Boilerplate can bloat projects quickly
- Requires careful performance tuning

React is the Clydesdale of the headless world: powerful, dependable, but not lightweight.

2. Vue: The Agile Workhorse

Best For: Developers who want a balance of power and simplicity.

Vue is elegant, approachable, and surprisingly versatile. Its single-file component structure makes it ideal for teams transitioning from more traditional development stacks.

- Smooth learning curve with a gentle ramp-up
- Great for mid-size projects and internal apps
- Excellent documentation and tooling (Vue CLI, Vite, Nuxt)
- Native support for transitions and reactivity

WordPress Tie-Ins:

- Pairs well with REST APIs—no need for GraphQL to get started
- Can be easily embedded in legacy WordPress templates for hybrid architectures

- Nuxt (Vue's full-stack framework) supports static generation and SSR

Downsides:

- Smaller ecosystem than React
- Not as commonly used in enterprise integrations (yet)

Vue is the thoroughbred: elegant, fast, and responsive—ideal for developers who like clean code and graceful performance.

3. Svelte: The Wild Mustang

Best For: Developers obsessed with performance and simplicity.

Svelte breaks with tradition by compiling components at build time. The result? Smaller bundles, blazing-fast rendering, and minimal runtime overhead.

- No virtual DOM = high-speed rendering
- Syntax is readable and minimal
- Strong support for reactive programming
- Paired with SvelteKit for routing and deployment

WordPress Tie-Ins:

- Fantastic for public-facing sites that fetch JSON content
- Clean for displaying Formidable Form entry data in small components or dashboards

- Can consume REST or GraphQL endpoints with minimal boilerplate

Downsides:

- Smaller community (but growing fast)
- Fewer enterprise-level integrations or starter kits
- Some advanced functionality still maturing (e.g., SSR edge cases)

Svelte is the Mustang—untamed, elegant, and lightning-quick. But it may buck the unwary rider.

Choosing the Right Ride for Your Stack

Framework	Best Use Case	Learning Curve	Ecosystem Size	Performance
React	Large apps, component libraries	Steep	Huge	Medium
Vue	Mid-size, readable projects	Gentle	Medium	High
Svelte	Performance-driven apps	Gentle	Small (growing)	Highest

---

Also consider:

- Your team's familiarity with JS tooling
- The desired performance profile of the application
- Ecosystem maturity (especially for plugins or design systems)

- Deployment preferences (e.g., Vercel vs Netlify vs custom CI)

Formidable Forms in Any Framework

No matter which frontend you choose, your Formidable Forms data remains accessible:

- Use fetch() or axios[1] to GET entries from /wp-json/frm/v2/entries
- POST new entries securely using token-based auth or a proxy endpoint
- Use repeaters, lookups, and calculated fields—logic handled server-side

Headless doesn't mean helpless. Formidable keeps your structured data safe and centralized.

Headless Wisdom

"Choose your mount wisely. For in battle, the wrong ride will throw you."

Takeaways:

- React, Vue, and Svelte are all valid choices for headless WordPress frontends.

- React offers maturity and scale; Vue provides balance and simplicity; Svelte offers speed and innovation.

- Formidable Forms integrates cleanly with all of them— your backend remains stable regardless of the frontend framework.

Next up: *"Formidable Without a Face: Using Formidable Forms in a Headless Setup"*—We'll go deep into the REST API structure, authentication, and how to build fully interactive headless workflows using Formidable.

¹What is Axios?

Axios is:

- **A promise-based HTTP client** for the browser and Node.js.

- Designed to simplify making HTTP requests (GET, POST, PUT, DELETE, etc.).

- Often preferred over fetch() because of:

- Automatic JSON transformation.

- Built-in support for request cancellation.

- Better error handling (treats non-2xx responses as errors).

- Support for interceptors to manipulate requests/responses.

Axios Example

```
import axios from 'axios';

axios.get('/wp-json/frm/v2/entries')
  .then(response => {
    console.log(response.data);
  })
  .catch(error => {
    console.error('Error fetching entries:', error);
  });
```

Whereas fetch() would require extra parsing:

```
fetch('/wp-json/frm/v2/entries')
  .then(response => response.json())
  .then(data => console.log(data))
  .catch(error => console.error('Error:', error));
```

Chapter 4

Formidable Without a Face: Using Formidable Forms in a Headless Setup

So far in our journey, we've explored the mythos of headless WordPress, dissected its architecture, and chosen a frontend steed worthy of enterprise deployment. Now we turn to the unsung hero of structured content and workflow management in this brave new world:

Formidable Forms—working silently behind the maskless UI.

In a headless setup, Formidable Forms becomes a true backend engine. It models data, enforces validation, runs calculations, stores entries, and exposes endpoints—all while remaining invisible to the end user.

Let's take a closer look at what happens when Formidable is decoupled from the frontend—and why it may be the best-kept secret in enterprise WordPress development.

⚜

What Does "Headless" Mean for Formidable Forms?

Normally, Formidable:

- Renders forms using shortcodes or Gutenberg blocks
- Displays data via Views and form HTML output
- Submits data through in-page AJAX or native POSTs

In a headless build:

- Forms are **not rendered by WordPress themes**— they're reconstructed in the frontend (e.g., React, Vue)
- Form submission is handled by **REST API calls**
- Data visualization is done with **custom frontend components**, not Formidable Views

But the **core logic remains**:

- Field validation still runs on submission
- Conditional logic is enforced
- Calculated fields are processed
- Entry storage is handled in frm_items and frm_item_metas
- Notifications, hooks, and actions still trigger as expected

Formidable becomes a **form modeling engine**, accessible via API.

Using the Formidable Forms REST API

To work with Formidable programmatically, you'll use its built-in REST API endpoints. Here are the key ones:

Get Form Structure

```
GET /wp-json/frm/v2/forms/{form_id}
```

This returns all fields, IDs, logic, and settings—perfect for dynamically building the frontend UI.

Submit an Entry

```
POST /wp-json/frm/v2/forms/{form_id}/entries
```

Request body must include:

```
{
  "item_meta": {
    "25": "John Doe",
    "26": "john@example.com"
  }
}
```

Include authentication headers or CSRF token (nonce) depending on your security model.

Fetch Entries

```
GET /wp-json/frm/v2/entries?form={form_id}
```

Query parameters can filter by user, field values, or custom conditions.

Frontend Integration Example (React)

Let's say your React app is rendering a contact form. You'd do something like:

```javascript
const submitForm = async (data) => {
  const response = await fetch('/wp-json/frm/v2/forms/12/entries', {
    method: 'POST',
    headers: {
      'Content-Type': 'application/json',
      'X-WP-Nonce': yourNonceVar
    },
    body: JSON.stringify({ item_meta: data })
  });
  const result = await response.json();
  // handle response
};
```

This decouples your interface completely from WordPress's presentation, while **still using Formidable's logic, validation, and entry management** under the hood.

Best Practices for Headless Formidable Use

1. **Model your forms in wp-admin.**
   Formidable still provides a fast UI for creating fieldsets, conditions, repeaters, and relationships.

2. **Avoid shortcode-based Views.**
   Handle data display with your frontend JS app using API calls.

3. **Use frm_after_create_entry and related hooks.**
   These still run on headless submissions—great for workflows, webhooks, or email logic.

4. **Secure the API endpoints.**
   Use JWT, OAuth, or WordPress nonce strategies

to validate calls. Do not expose form logic blindly to the public.

5. **Cache form structure.**
   If you fetch field definitions to build UIs, cache them locally or with a serverless layer to reduce load.

6. **Treat Formidable as a data-entry microservice.**
   It's now your gateway to user input—not your UI layer.

Advanced Use Case: Approval Workflows

In an enterprise workflow system, you might:

- Build a multi-page intake form using Formidable
- Submit entries from a Vue frontend to the API
- Run backend logic to trigger conditional approval steps
- Fetch role-based entries into a React dashboard
- Apply custom logic with frm_after_update_entry to send notifications or transition states

You've now created a **workflow engine** inside WordPress, using Formidable as your structured data core—**without rendering a single WordPress page.**

<hr/>

Headless Wisdom

"The form does not need a face to shape the data—it only needs structure."

<hr/>

Takeaways:

- Formidable Forms works seamlessly in headless architecture via its REST API
- All entry logic and validation remains functional without rendering the form in WordPress
- Use Formidable for secure, structured data collection while your frontend handles UX

<hr/>

Next up: *"Building the Bridge: Authentication, SSO, and Secure API Gateways"*—We'll look at how to secure headless WordPress applications with modern auth flows—like Sign in with Google, token validation, and SSO integration.

Chapter 5

Building the Bridge: Authentication, SSO, and Secure API Gateways

Let's imagine for a moment: the Headless Horseman is galloping through a digital forest. He's fast, formidable, and terrifyingly efficient. But unless someone controls the **bridge**, he can cross into places he shouldn't—like your sensitive data.

In the world of headless WordPress, that bridge is your **authentication system**. And in this article, we're fortifying it with modern best practices for:

- Single Sign-On (SSO)

- REST API security

- Token-based authentication

- Custom access control for Formidable Forms data

Let's get our horseman over the bridge—without leaving the gate open behind him.

Why Authentication and API Security Matter in Headless WordPress

In a traditional WordPress site, authentication is mostly session-based. But once you decouple the frontend, you're in API territory. Now you need to:

- Authenticate external clients (React, mobile apps, other servers)

- Authorize user-specific access to entries and forms

- Protect sensitive or PII data from being exposed via open REST routes

Authentication Options in a Headless WordPress Stack

There are multiple ways to authenticate users with a headless frontend:

1. Cookie Authentication

- Uses WordPress's native login cookies
- Simple, but only works within the same domain and browser

2. Application Passwords

- Each user generates a token-like app password
- Passed via HTTP Basic Auth headers
- Easy to implement, limited in granularity

3. JSON Web Tokens (JWT)

- Stateless and portable
- Enables decoupled clients (e.g., React) to authenticate securely
- Requires a plugin or custom logic to issue and validate tokens

4. OAuth2 for SSO

- Enterprise-grade
- Integrates with providers like Google, Apple, Auth0, or Azure AD
- Enables federated identity across systems

Securing REST Endpoints: Replace Filters with Real Control

Formidable Forms does not provide a native hook for managing REST-level access control, but its extensible architecture makes it possible to implement secure, role-based authentication—especially when paired with custom API permissions or external SSO gateways.

WordPress provides a powerful mechanism:

```php
add_action('rest_api_init', function () {
  register_rest_route('my-namespace/v1', '/entries/(?P<id>\d+)', [
    'methods'  => 'GET',
    'callback' => 'my_custom_get_entry',
    'permission_callback' => 'my_custom_entry_permission_check',
  ]);
});

function my_custom_get_entry($request) {
  $entry_id = $request['id'];
  $entry = FrmEntry::getOne($entry_id, true);

  if (!$entry) {
    return new WP_Error('not_found', 'Entry not found', ['status' => 404]);
  }

  return [
    'id'   => $entry->id,
    'meta' => $entry->metas,
  ];
}

function my_custom_entry_permission_check($request) {
  $current_user = wp_get_current_user();

  // Replace with your actual role/capability logic
  if (!current_user_can('read_private_posts')) {
    return new WP_Error('forbidden', 'You do not have permission to access
  }

  return true;
}
```

This approach provides the control that the missing Formidable hook cannot. You own the route. You gatekeep the access.

⚜

Verifying Identity: Token Flow and Reverse Proxies

Once SSO or token-based logins are introduced, the identity token must be verified:

- JWT tokens must be decoded and validated on the server
- OAuth tokens should be checked against the issuer's introspection endpoint
- Reverse proxies like **Kong**, **Amazon API Gateway**, or **Cloudflare** can handle token validation before WordPress is even touched

This layered approach minimizes risk and aligns with modern zero-trust models.

Headless Wisdom

"Security isn't a feature—it's a design decision made early or paid for later."

Takeaways:

- Enterprise-grade authentication starts with intentional system design, not patchwork fixes.

- WordPress can support secure SSO and API authentication when built around proper REST practices.

- Avoid relying on undocumented or nonexistent plugin hooks—always verify the integration points.

- permission_callback in register_rest_route() gives you full control over who sees what.

- Gateways, middleware, and token-based identity checks form the bedrock of scalable, secure APIs.

⚔

Next up: *"The Horseman's Trail: Deployment Models for Headless WordPress"*—We'll explore how to host, scale, and deploy your frontend and backend pieces—plus tips for staging, CI/CD, and enterprise testing workflows.

Interlude: What Is CI/CD?

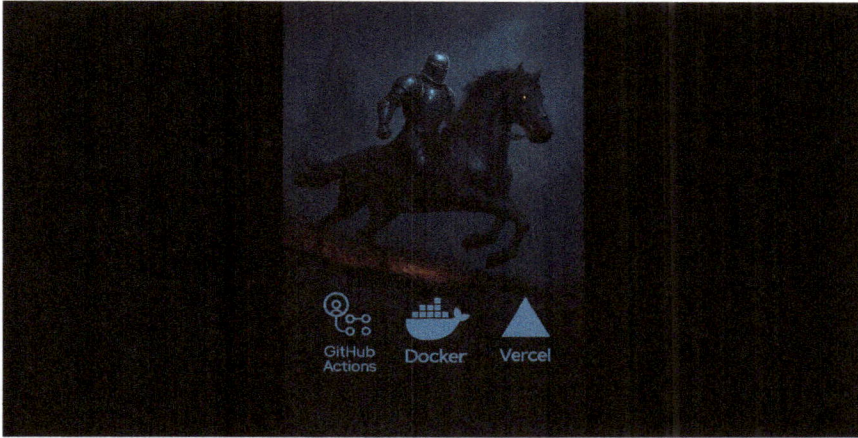

Imagine the Headless Horseman had to stop and lace up his boots every time he reached a village. It wouldn't be terrifying—it would be tedious. Now apply that metaphor to deploying code. In enterprise-grade headless systems, **manual deployment is the enemy of speed, scale, and sanity.**

That's where **CI/CD**—Continuous Integration and Continuous Deployment—rides in like cavalry. In this article, we'll explore how to implement modern automation pipelines for headless WordPress architectures built on Formidable Forms, REST APIs, and JavaScript frontends.

Because nothing says enterprise maturity like repeatable, testable, and automated releases.

⚔

What Is CI/CD?

Continuous Integration (CI)

CI is the practice of automatically building and testing your application whenever code is pushed to the repository. It ensures:

- Code quality is maintained
- Bugs are caught early
- Team collaboration is frictionless

Continuous Deployment (CD)

CD takes the validated code and automatically pushes it to staging or production environments. This can include:

- Static site builds (e.g., React, Vue)
- WordPress theme/plugin updates
- Asset compilation
- Cache purging and CDN invalidation

Together, CI/CD bridges the gap between code and production, safely and swiftly.

CI/CD in a Headless WordPress Stack

A headless architecture introduces unique challenges and opportunities for automation. Your stack might look like this:

- **Frontend**: React or Vue
- **Backend**: WordPress with Formidable Forms
- **API**: REST or GraphQL endpoints
- **Infrastructure**: GitHub + Vercel/Netlify + Kinsta/WPEngine

Let's look at how CI/CD plays a role across that stack.

Frontend CI/CD Flow

1. Developer pushes to GitHub (e.g., main or dev)
2. GitHub Actions runs:
   - ESLint/Prettier
   - Unit/Integration tests (e.g., Jest, Cypress)
   - Static builds via npm run build
1. Deployed to Vercel/Netlify

2. Preview URLs generated for stakeholders

Backend CI/CD Flow

1. Plugin or theme changes pushed to GitHub

2. Composer runs (if used)

3. WP-CLI automates:

   o Plugin activation

   o Formidable form import

   o Option syncing

4. Deployment to WordPress host (e.g., Kinsta, WP Engine) via SSH, FTP, or Git

5. Optional: run post-deploy tests or warm caches

Tooling for CI/CD

Here are some enterprise-ready tools for building your automation bridge:

Layer	Tools
CI Orchestration	GitHub Actions, GitLab CI
Frontend Hosting	Vercel, Netlify, AWS Amplify
WordPress Sync	WP-CLI, Composer, SFTP Deploy
Containerization	Docker, Local by Flywheel
Secrets & Vaults	GitHub Secrets, Doppler, Vault
Testing	PHPUnit, Cypress, Playwright

For Formidable Forms, pair WP-CLI with JSON-formatted form exports and version them in Git. It brings schema discipline to the otherwise click-happy admin panel.

Security in the Pipeline

CI/CD isn't just about speed—it's also about governance:

- Rotate and secure deploy keys
- Use GitHub Actions with environment protection rules
- Include security scanning in your pipeline (e.g., npm audit, composer audit)
- Set up rollback strategies or blue/green deployments

Headless Wisdom

"A headless system is only as agile as its deployment pipeline."

Takeaways:

- CI/CD = Continuous Integration + Continuous Deployment

- Use CI to enforce quality with linting, testing, and builds

- Use CD to automatically deploy React frontends and WordPress plugins/themes

- WP-CLI + Git = version-controlled Formidable Forms

- Secrets management and rollback strategies are non-negotiable in enterprise stacks

- CI/CD is not just for big tech—it's essential for headless WordPress maturity

Chapter 6

The Horseman's Trail: Deployment Models for Headless WordPress

The frontend is decoupled. The backend is WordPress. Formidable Forms is humming along beneath the surface. So the question now becomes...

Where does the Horseman ride?

Deployment is the unsung hero of headless WordPress success. A brilliant architecture can collapse under the weight of a poorly planned deployment pipeline. In this chapter, we'll explore deployment models for both frontend and backend,

45

CI/CD workflows, and infrastructure strategies for scaling enterprise-grade headless WordPress systems.

⚔

A Tale of Two Worlds: Frontend and Backend Deployment

In headless builds, you're dealing with two separate deployable systems:

Layer	Hosted On	Typical Tools
Frontend	Jamstack/CDN edge	Vercel, Netlify, Cloudflare Pages
Backend	PHP server	WP Engine, Kinsta, SpinupWP, Docker

Each has its own environment, build triggers, caching behavior, and deployment timeline.

⚔

Frontend Deployment Options

Static Site Generation (SSG)

Tools like **Next.js, Nuxt**, and **SvelteKit** can pre-render content during the build phase.

- **Pros:** Blazing fast, CDN-cached, SEO-friendly
- **Cons:** Requires re-builds when content changes

46

- **Best For:** Marketing sites, blogs, or infrequent content changes

Tip: Use a webhook from WordPress to trigger a rebuild in Vercel/Netlify when entries are added via Formidable Forms.

Server-Side Rendering (SSR)

Your JavaScript app queries the WordPress REST API in real time on page load (e.g., /blog fetches entries dynamically).

- **Pros:** Always up-to-date, no rebuilds needed
- **Cons:** Slower TTFB, requires secure API handling
- **Best For:** Authenticated dashboards, internal portals, real-time content

Incremental Static Regeneration (ISR)

Hybrid model (available in Next.js) that lets you statically generate most content, then **revalidate** after a timeout.

- **Best For:** Sites with a lot of semi-static content
- **Example:** Cache form submissions, but revalidate the form index page every 60 seconds.

Backend Deployment Options (WordPress + Formidable)

Headless doesn't mean you abandon WordPress best practices.

Managed WordPress Hosting (Fastest Setup)

Services like WP Engine or Kinsta are ideal for stable deployments:

- Automatic backups
- Easy staging environments
- Redis + object cache options

Pair with Git-based deploys via GitHub Actions or GitLab CI for plugin/theme changes.

Dockerized WordPress (Customizable & Scalable)

Spin up your own stack using:

- Docker + Docker Compose

- Bedrock or custom WP scaffolding

- MariaDB/Postgres + Redis

- Nginx or Traefik + Let's Encrypt

Perfect for agencies or enterprise teams needing version-controlled infrastructure.

⊶⊷✝⊶⊷

Serverless Hybrid (API-as-a-Service)

You can abstract away the frontend even more:

- Deploy the frontend to Vercel/Cloudflare

- Host only the WP REST API on serverless WordPress (e.g., WP Cloudflare Worker proxy)

- Formidable Forms stores and returns data via fetch calls

- This is cutting-edge—but not for the faint of heart.

⊶⊷✝⊶⊷

CI/CD for the Headless Stack

Treat your frontend and backend as separate repos with independent pipelines.

Concern	Frontend Pipeline	Backend Pipeline
Codebase	JS (React/Vue/Svelte)	PHP (Theme/Plugin/Bedrock)
Hosting	Vercel, Netlify, Cloudflare	WP Engine, Kinsta, Docker VPS
Deployment trigger	Push to main or webhook	Push to main or tag release
Testing	Cypress, Playwright	PHPUnit, WP CLI, custom scripts

Recommended CI tools:

- GitHub Actions
- Bitbucket Pipelines
- Buddy
- DeployHQ

Optional: Add **content webhook triggers** when WordPress changes occur (e.g., Formidable entry submitted → rebuild form archive page).

⚔

Formidable Forms Deployment Considerations

- **Keep Form Definitions in Sync**
    - Export/import form JSON as part of deployment or staging push.
- **API Tokens and Webhooks**
    - Ensure secure keys (e.g., Google Maps, Stripe) are environment-specific.
- **Entry Data**
    - Never move entries between environments. Use a plugin like WP Migrate Lite to clone only the schema.
- **Entry Hook Logic**
    - Validate that hooks like frm_after_create_entry don't fire unintentionally during imports.

⚔

Headless Wisdom

"It's not the blade that wins the battle, but the one who knows how to sharpen and wield it."

Takeaways:

- Deploy frontend and backend separately with tools suited to each layer
- Choose SSG, SSR, or ISR depending on how dynamic your data is
- Use CI/CD to automate, test, and scale your deployments
- Formidable Forms works cleanly across environments with API-first architecture

Next up: *"When the Horse Stumbles: Debugging and Monitoring a Headless Stack"*—We'll dig into error tracing, logging, and observability techniques for when the data doesn't show up... or shows up sideways.

Chapter 7

When the Horse Stumbles: Debugging and Monitoring a Headless Stack

A headless architecture may gallop like a champion—but even the strongest horse stumbles.

Debugging in a traditional WordPress site usually means popping open the browser dev tools, checking the error logs, or disabling a plugin. But in a **headless WordPress environment**, problems often hide in the gaps between layers: WordPress backend, REST API, JavaScript frontend, or deployment pipeline.

This article is your guide to finding and fixing the issues that can break headless systems—and keeping your Horseman upright, operational, and monitored in real time.

The Headless Debugging Landscape

In a headless stack, your debugging toolkit must span at least four distinct domains:

Layer	Common Failures
WordPress Backend	Plugin conflicts, database schema issues
REST API Layer	Permissions, malformed responses, CORS
JavaScript Frontend	Fetch errors, rendering issues, state bugs
Deployment/CI	Build failures, stale data, cache mismatch

Each layer speaks its own language. To master the stack, you must translate across them.

Diagnosing WordPress API Issues

Start here when the frontend can't load data.

Enable WP_DEBUG

```
define('WP_DEBUG', true);
define('WP_DEBUG_LOG', true);
```

Then check wp-content/debug.log after reproducing the issue.

- **Use Postman or Insomnia**
  - Test your /wp-json/ endpoints directly. Does a 403 show up? 500? CORS block?

- **Check Permission Callbacks**
    - o If using custom REST routes, ensure permission_callback returns true or proper capability checks.

- **Validate Formidable Responses**
    - o Sometimes item_meta arrays or calculated fields silently fail. Use the frm_entries_before_create filter to log values.

JavaScript Frontend Debugging

The frontend often appears broken when it's just missing data.

- **Use Browser Dev Tools**
- Check network tab for REST responses
- Look for CORS errors (e.g., "No 'Access-Control-Allow-Origin' header")
- **Log API Failures**

```
try {
  const res = await fetch('/wp-json/...');
  if (!res.ok) throw new Error(`Status: ${res.status}`);
  const data = await res.json();
} catch (err) {
  console.error('Fetch failed:', err);
}
```

- **Validate Nonce/JWT Headers**
    - o If using X-WP-Nonce, make sure it matches the user session

- If using JWT, verify token expiration and audience

Cache and Deployment Gotchas

- **Purge CDN Cache**
  - If you're using Vercel or Netlify, make sure rebuilds are triggered after content updates. Formidable Form submissions don't automatically refresh cached pages.

- **Stale API Responses**
  - Use server-side rendering or client-side revalidation to avoid relying on outdated entry data.

- **Debug CI Failures**
  - If builds are breaking:
    - Review commit diffs
    - Ensure environmental variables (like API base URLs) are correctly scoped
    - Use verbose logs in GitHub Actions or Netlify deploy logs

Monitoring Tools for a Headless Stack

Tool	Use Case
Sentry	Frontend JavaScript error logging
LogRocket	Full-session replay of user interactions
WP Activity Log	Backend user/action tracking in WordPress
New Relic / Datadog	APM and performance monitoring
Upptime / Pingdom	Uptime and endpoint health

Formidable Forms Debugging Tips

- Use the **Formidable Logs tab** if enabled, especially for email and webhook troubleshooting.

- Attach custom logging to filters:

```
add_filter('frm_entries_before_create', function($errors, $form){
    error_log('Form submission received: ' . print_r($_POST, true));
    return $errors;
}, 10, 2);
```

- Check that field keys are **not changed** between staging and production—especially for dynamic or calculated fields.

Headless Wisdom

"The stumble is not the end of the journey—only a chance to learn where the ground was weak."

Takeaways:

- Debugging headless WordPress requires tools across API, frontend, and infrastructure layers
- Monitor REST permissions, CORS, and data shape carefully
- Formidable Forms can be fully logged and monitored through WordPress hooks

Next up: "*The Townsfolk React: Client Education and Stakeholder Buy-In*"—We'll discuss how to present the value of headless WordPress to non-technical clients and stakeholders—and how to get early buy-in for decoupled architectures.

Chapter 8

The Townsfolk React: Client Education and Stakeholder Buy-In

The Horseman rides confidently—but the townsfolk are skeptical.

You've designed a decoupled architecture. You've deployed WordPress as a backend engine, Formidable Forms as a structured data core, and React as your frontend of choice. The tech team is thrilled.

But what about the people who write the checks?

"Why doesn't it look like WordPress anymore?"

"Why can't I drag and drop my pages?"
"Did we just break the whole website?"

This chapter is your guide to **client education, executive reassurance, and stakeholder buy-in**. Because headless systems don't just change the tech—they change expectations.

⁂

Understanding Stakeholder Concerns

In headless WordPress projects, resistance is usually not about the technology—it's about **visibility, control, and cost**.

Common Concerns:

- *"Will our content team still know how to use it?"*
- *"Why is this more expensive than a plugin?"*
- *"Can we still use Yoast SEO, page builders, or shortcodes?"*
- *"What happens if the API breaks?"*

All valid. And all solvable with clear communication.

⁂

Translate Headless into Business Benefits

You need to speak in **outcomes**, not architecture. Here's how to reframe the conversation:

59

Technical Change	Stakeholder-Friendly Benefit
Decoupling frontend	Faster performance, scalable design, multichannel delivery
REST API instead of pages	Integrates with future systems (CRM, LMS, mobile, etc.)
Formidable as data engine	No need to rebuild forms or workflows
CI/CD pipelines	Faster iteration, better rollback and testing
Custom frontend (React/Vue)	Fully branded, UX-optimized UI

Example pitch:

"We're replacing the theme with a lightning-fast frontend that loads instantly, works better on mobile, and can be reused across your web app, your internal dashboard, and even your future mobile experience."

That's a far better sell than:

"We removed the template system and are rendering asynchronously from a decoupled JS stack."

Offer a Guided Content Authoring Experience

Non-technical users still need a comfortable way to:

- Enter structured data
- Edit content
- View submissions and workflows

Formidable Forms makes this possible.

- Use the WordPress admin panel as the internal CMS
- Lock down confusing menu items
- Build structured entry forms for content types
- Offer preview modes in the frontend app

Pro tip: Include short admin screencasts as part of your deliverables—this builds confidence and reduces support overhead.

Introduce the "Progressive Decapitation" Strategy

If full decoupling feels too sudden, suggest a phased approach:

1. Keep the main site WordPress-powered, but build one React-powered dashboard
2. Use Formidable Forms for both systems, creating cross-compatible data
3. Gradually migrate public pages to the frontend
4. Reuse the backend CMS with new UIs over time

This reduces fear and allows stakeholders to *see* the value before committing fully.

Tools That Build Confidence

- **Loom or Bubbles** – Walkthroughs of new workflows
- **Miro or FigJam** – Visual diagrams for data flow and content model
- **Notion or Confluence** – Living documentation hub for how it all works
- **Slack Channels** – Dedicated support thread with fast responses early on

What to Say (and What Not to Say)

Say:

- "Your team will still use WordPress to manage content."
- "We're replacing the theme, not the whole platform."
- "This structure is future-proof—it gives you API access to your own content."

Don't Say:

- "It's decoupled. You'll figure it out."
- "The frontend isn't WordPress anymore, so good luck."
- "We just made it a JSON firehose and dropped the UI."

Even developers need to remember: *you're not selling technology—you're selling clarity*

⚜

Headless Wisdom

"When the village understands the vision, the rider is never alone."

⚜

Takeaways:

- Headless systems require education—not just engineering
- Reframe technical features as business benefits
- Formidable Forms enables a familiar content interface inside a decoupled world
- Use visuals, demos, and support to build trust early and often

⚜

Next up: *"Ghosts in the Code: Legacy Plugins, Compatibility, and Gotchas"*—We'll uncover what breaks, what survives, and what needs replacing when you move WordPress into headless territory.

Chapter 9

Ghosts in the Code: Legacy Plugins, Compatibility, and Gotchas

As the Horseman thunders through modern APIs and frontend frameworks, he occasionally rides past graveyards—quiet places where old plugins linger and legacy systems whisper warnings.

Going headless is powerful, but not without trade-offs. Some of the magic that makes WordPress user-friendly—shortcodes, theme-based rendering, plugin UIs—doesn't survive the decoupling process.

In this article, we'll identify what breaks, what adapts, and what to avoid when transitioning a WordPress site to headless architecture—especially when Formidable Forms is your structured data engine.

What Breaks in a Headless WordPress Build

Let's start with the rough news. These features are either **fully or partially incompatible** when you decouple the frontend:

Shortcodes

- Shortcodes won't render in your React/Vue/Svelte frontend.
- You'll need to replicate their behavior manually—or replace them entirely.

Page Builders (Elementor, WPBakery, Divi)

- These tools depend on the WordPress theme layer.
- The rendered output is lost when you bypass the_content().

Widgets

- Widgets are tied to theme regions like sidebars or footers.
- A headless frontend has no concept of widget zones.

Menus Built in PHP

- wp_nav_menu() won't render. You must query menus via the REST API or rebuild them in your frontend app.

What Still Works—But Needs Workarounds

Yoast SEO / Rank Math

- These plugins output meta tags to the page header.
- You must extract SEO fields via REST and inject them manually into your frontend (next/head, <Helmet>, etc.).

Contact Forms (Non-API-Based)

- Gravity Forms, Contact Form 7, and others may rely on shortcode rendering and page refreshes.
- If you're not using Formidable, you'll need to POST via admin-ajax or REST (if supported).

Customizer

- The WordPress Customizer is theme-centric.
- In headless, you configure styles and layouts in your frontend code, not via the admin panel.

Formidable Forms: Ghost-Free Integration

Here's where Formidable Forms shines:

REST API Ready

- Fetch form structures, post entries, retrieve data, and run logic without rendering a single WordPress page.

No Theme Dependencies

- All logic lives in backend hooks (frm_after_create_entry, frm_entries_before_create, etc.)

Structured Data First

- Ideal for headless projects that need custom fieldsets, repeaters, file uploads, and calculated fields.

Entry Management UI Remains Intact

- Admins still view entries, stats, and workflows in /wp-admin.

Programmatic Customization is Clean

- Use PHP filters to modify data, validate fields, and format API output without breaking decoupling.

Plugins That Work Well with Headless WordPress

Plugin Category	Compatible Options
SEO Fields	Yoast / Rank Math (extract via REST)
Forms / Data	Formidable Forms, WPForms (API editions)
ACF / CPT UI	Both work great for backend modeling
WPGraphQL	Ideal for GraphQL consumers
WP REST Cache	Speed up API responses
WP JWT Auth / OAuth	For secure API authentication
WP Activity Log	Admin-side audit trail

Migration Tips: Avoiding the Haunted Forest

1. **Audit Your Plugins**

   Before decoupling, list all active plugins and test their functionality via REST or frontend replacement.

2. **Shortcodes Early**

   Replace shortcode-based layouts with custom blocks, JSON-driven components, or structured entry data.

3. **API-Only Philosophy**

   Favor plugins with clear REST (or GraphQL) endpoints, not those that output directly to the_content.

4. **Document Workarounds**

   Keep a changelog of plugins or features replaced during decoupling. It helps with onboarding and support.

5. **Build "Headless-First" from Day One**

New features should rely on API availability—not template injection.

⚔

Headless Wisdom

"You can't ride into the future dragging the skeletons of the past."

⚔

Takeaways:

- Many legacy plugins break or become irrelevant in a headless architecture
- Formidable Forms is API-driven and built for headless success
- Avoid shortcode-heavy tools, theme-dependent plugins, and UI-bound logic
- Plan your migration carefully and document everything

⚔

Next up: *"The Headless Future: Where This Is Going and Why It Matters"*—We'll close the series by looking at trends in composable architecture, low-code platforms, and why structured WordPress + JavaScript may be the best of both worlds.

Chapter 10

The Headless Future: Where This Is Going and Why It Matters

The Horseman doesn't just haunt the present—he rides toward the future, fast and unswerving. But where is this trail headed?

In our final article of the series, we set our eyes on the road ahead: a world of composable architecture, low-code platforms, and structured data ecosystems where WordPress—yes, WordPress—can still lead, if we ride wisely.

The Composable Web: Breaking Down the Monolith

Traditional CMSs like WordPress bundle everything: content editing, page rendering, theming, and data logic. But the future is **composable**—where every piece of the stack does one thing extremely well and communicates through APIs.

- **Content from WordPress**
- **Forms and data from Formidable**
- **Search from Algolia**
- **Images from Cloudinary**
- **Frontend from React/Next.js**

The result? Flexibility, scalability, and a future-proof ecosystem. It's not just JAMstack—it's the unstacking of assumptions.

Structured Data Is the New Content

Blogs made WordPress popular. Pages and posts still matter. But in enterprise systems, **structured data**—not prose—is king.

Think:

- Patient records
- Grant applications

- Inventory data
- Workflow approvals

Formidable Forms gives WordPress a shot at competing with Salesforce, Airtable, and even purpose-built low-code platforms. Combine it with a frontend that consumes JSON via REST, and you're not building a website—you're architecting a system.

<hr />

Trends Driving Headless Adoption

1. **Performance & Core Web Vitals**

   Lighthouse scores matter. Headless setups with static prerendering blow past monolithic WordPress on speed.

2. **Developer Experience**

   Teams prefer JavaScript frontends, component libraries, and IDE-native tooling. Headless enables modern DX.

3. **Security Hardening**

   Fewer public-facing PHP endpoints = reduced attack surface.

4. **Scalable Multisite and Multi-tenant**

   API-first setups let you plug the same data source into multiple frontends: mobile apps, portals, microsites.

5. **AI and Workflow Automation**

Structured data in headless CMSs integrates more cleanly with AI models and automation pipelines.

And Yet... WordPress Isn't Going Away

WordPress still runs **over 40% of the web**. It's familiar, extensible, and backed by a massive plugin ecosystem. The key is **knowing when to decouple**—and how to do so without breaking what users love.

Headless isn't about abandoning WordPress. It's about elevating it beyond its origin story.

Formidable Forms helps you do that. It bridges legacy and modern, forms and fields, admin and API.

So, Where Do We Go From Here?

As developers and architects, we must:

- Advocate for structured data
- Design frontend-first but backend-aware
- Choose tools that scale with clients' ambitions

- Keep learning—because the horseman never stops riding

$$\dagger$$

Thanks for reading. May your APIs stay restful, your deployments stay atomic, and your Horseman ride far.

About the Author

Victor M. Font Jr. is a seasoned enterprise consultant, full-stack WordPress developer, and the founder of Formidable Masterminds—a platform empowering developers to build structured, scalable applications with Formidable Forms. With decades of experience spanning business intelligence, systems integration, and software architecture, Victor brings a uniquely pragmatic approach to WordPress as a serious application platform.

He is also the creator of *Developers Corner*, a thought leadership hub where technical depth meets developer education. Victor's work bridges the gap between enterprise software practices and the WordPress ecosystem, equipping professionals with the tools and mindset to build what matters.

When he's not wrangling REST endpoints or designing scalable data models, Victor enjoys writing, mentoring, and decoding the theology of technology.

Call to Action / Next Steps

If this book helped shift your perspective on what's possible with WordPress and Formidable Forms, here's how to go further:

- **Join the Developers Corner community**

  - Visit [Formidable Masterminds Facebook Group](#) to discover premium add-ons, tutorials, and enterprise-grade code snippets built for Formidable Forms power users.

- **Subscribe for future releases**

  - Stay ahead of the curve with in-depth articles, code breakdowns, and announcements on headless strategies and structured data workflows. Visit [formidable-masterminds.com](#) to subscribe.

- **Follow Victor's thought leadership**

  - On LinkedIn, X, and WordPress circles, you'll find Victor sharing insights that challenge, inspire, and guide developers building real-world systems—not just websites.

You're not just building a frontend. You're engineering a

platform. Let's build better, together.

Headless Reference Materials

Glossary of Terms

API (Application Programming Interface)

A set of defined rules that enable software applications to communicate with each other. In WordPress, REST APIs allow external systems to read or write data to the backend.

Axios

A promise-based JavaScript HTTP client for the browser and Node.js, commonly used in frontend frameworks like React and Vue. Axios simplifies API requests and provides built-in support for request cancellation, automatic JSON parsing, interceptors, and more robust error handling than the native fetch() API.

CI/CD (Continuous Integration / Continuous Deployment)

A development practice that automates the process of integrating code changes and deploying them to production.

CI ensures code is tested and validated continuously, while CD automates the release process, reducing deployment risk.

Composable Architecture

A modular design strategy where each system component is API-driven and independently maintainable.

FaaS (Functions as a Service)

A serverless computing model where developers deploy individual functions that run in response to events. Popularized by platforms like AWS Lambda, FaaS enables scalable, event-driven backends without managing servers.

Formidable Forms

A powerful WordPress form builder plugin that enables developers to create complex forms, structured data models, and dynamic front-end views. In a headless architecture, it can serve as the structured data layer.

Frontend

The user-facing portion of an application, typically built with HTML, CSS, and JavaScript. In a headless WordPress setup, the frontend is decoupled and built using frameworks like React or Vue.

Headless CMS

A content management system that provides data via API without handling presentation or frontend rendering. WordPress can be used as a headless CMS by exposing content through the REST API.

Hydration

The process of rendering server-generated HTML with JavaScript interactivity on the client side.

JWT

JSON Web Token, a compact, self-contained token used for secure API authentication.

OAuth2

An open standard for access delegation. It allows applications to authenticate users and grant access without exposing passwords, enabling secure SSO (Single Sign-On) workflows.

REST (Representational State Transfer)

An architectural style for designing networked applications. WordPress's REST API conforms to REST principles, allowing external clients to interact with WordPress content via HTTP requests.

SSO (Single Sign-On)

A user authentication process that allows users to log in once and gain access to multiple systems without logging in again. Commonly used in enterprise systems via OAuth2 or SAML.

Structured Data

Information organized in a predictable and machine-readable format. In WordPress, this can be implemented using custom post types, taxonomies, and meta fields—or with form builders like Formidable Forms.

Recommended Resources

Formidable Forms Developer Docs

Formidable REST API Docs

WP REST API Handbook

Headless WordPress Stack Explorer

Next.js Documentation

Plugin Compatibility Reference

Plugin Type	Compatible in Headless?	Notes
Formidable Forms	Yes	Native API, ideal for structured data
Classic Editor	Limited	No frontend output
WPBakery	No	Renders only in PHP, not API-based
ACF (REST exposed)	Conditional	REST support needed for fields
Gutenberg Blocks	Partial	Needs hydration in JS frontend

Coming Soon from Formidable Masterminds Press

Continue your journey toward enterprise mastery with our upcoming developer guides, each designed to expand your capabilities in real-world WordPress development.

Relational WordPress: Database Design and Performance for Formidable Developers

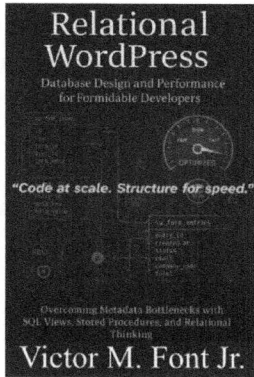

Arriving November 2025

Push past the limits of metadata. Learn to extract, normalize, and optimize Formidable Forms data using SQL views, stored procedures, and custom reporting schemas.

Formidable Workflows: Automating Business Processes with WordPress

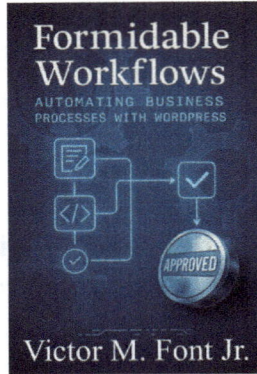

Arriving February 2026

Model your client's business rules using Formidable's actions, views, and conditional logic. Learn how to turn forms into fully automated applications—without writing custom plugins.

Securing the Stack: Identity and Access in WordPress Applications

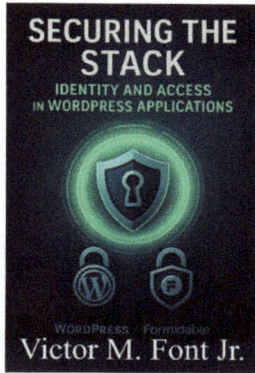

Arriving May 2026

Gain confidence in protecting your systems. Explore OAuth2, SSO, token-based authentication, and custom access logic—tailored for headless and hybrid WordPress architectures.

Composable WordPress: Modular Systems with Custom Plugins and APIs

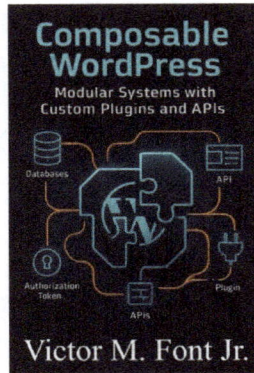

Arriving August 2026

Unlock a modular future with custom plugins, REST endpoints, and decoupled frontends. Learn how to think in components while keeping WordPress as your orchestrator.

The Formidable Developer's Playbook: Hooks, Filters, and API Mastery

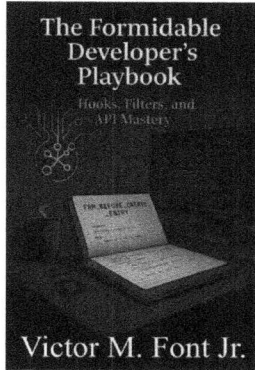

Arriving November 2026

Master the Formidable Forms developer lifecycle. Build clean, extensible add-ons using the right filters, actions, and REST strategies—direct from the code trenches.

Join us at FormidableMasterminds.com
for early access, downloadable code, and bonus material from the series.

www.ingramcontent.com/pod-product-compliance
Lightning Source LLC
Chambersburg PA
CBHW060633210326
41520CB00010B/1585